SW

The Water Shrew Handbook

by Phoebe Carter and Sara Churchfield

Contents

The Water Shrew Handbook
by Phoebe Carter and Sara Churchfield

Published by The Mammal Society

The Mammal Society
Registered Charity No. 278918
Registered Office:
The Mammal Society
2B Inworth Street
London SW11 3EP

ISBN 0 906282 62 4

This is one of a series of booklets on British mammals published by The Mammal Society

The Mammal Society is extremely grateful to the Environment Agency's Science Programme for sponsoring the publication of this book and for helping to fund The Mammal Society's Water Shrew Survey (2004-2005).

Photographs: Rollin Verlinde ©Vilda (www.vildaphoto.net) Cover & Figs. 4, 5, Yves Adams ©Vilda (www.vildaphoto.net) Fig. 1, Alasdair Love Fig. 2, Martin Smith Fig. 3, David Hosking Figs. 6, 7, Sara Churchfield Figs. 8, 14, 15, 18 & all in Appendix 2, Angela Gall Figs. 12, 16, Dr Steve Furness Fig. 13, Phoebe Carter Fig. 17.

Layout and printing by SP Press, Units 1 & 2, Mendip Vale Trading Estate, Cheddar Business Park, Wedmore Road, Cheddar, Somerset BS27 3EL.

1 Introduction

The water shrew is one of Britain's least known but most intriguing mammals. The elusive nature and discreet field signs of the water shrew make it a difficult species to study and, consequently, it has been overlooked in comparison with other riparian species. Occupying similar habitats to the water vole, it was thought that the water shrew may have suffered a similar fate, with changes in land-use and habitat management contributing to a decline in its numbers and occurrence. There has also been concern that pollution and pesticide use may have had an adverse effect on this species.

Intensive, localised studies generated some of the first information on the behaviour and diet of water shrews, and on their habitat occurrence (Churchfield, 1984a,b, 1985, 1990, 1991, 1998; French *et al.*, 2001; Greenwood *et al.*, 2002). Following these studies and guided by the recommendations in the Species Action Plan for the water shrew (Churchfield, 1997), The Mammal Society initiated a two-year, nationwide Water Shrew Survey (2004-2005). This first national dataset on the distribution and habitat occurrence of water shrews has provided a baseline against which future changes to their populations can be assessed. While greatly increasing our knowledge of water shrews, these studies are just the start of the work that is needed to assess the conservation needs of this charismatic species fully.

1.1 Purpose of The Water Shrew Handbook

The aim of this Handbook is to raise awareness about water shrews. It includes information on their biology and ecology, provides guidance on carrying out surveys for water shrews and makes suggestions as to how we can ensure their continued survival. It is hoped that the production of this Handbook will lead to the water shrew becoming better studied and recorded, and more highly valued as an important part of the riparian ecosystem.

2 Water shrew biology and ecology

2.1 Taxonomy

The water shrew, *Neomys fodiens* (Pennant, 1771), belongs to the order Soricomorpha. Shrews, moles and hedgehogs were previously assigned to the order Insectivora, together with other small, ground-dwelling insectivorous mammals such as tenrecs, golden moles and solenodons. Accumulating anatomical and molecular research is providing compelling evidence that these groups of mammals do not share a common line of descent and so they have been re-assigned to better reflect their disparate phylogenetic relationships (see Wilson & Reeder, 2005). Accordingly, all species of shrew, along with the talpid moles, are assigned to a distinct order, Soricomorpha.

The five other shrews in this order that are found in the British Isles are the common shrew (*Sorex araneus*), the pygmy shrew (*Sorex minutus*), and on some of the Channel and Scilly islands, the lesser white-toothed shrew (*Crocidura suaveolens*), the greater white-toothed shrew (*Crocidura russula*) and Millet's shrew (*Sorex coronatus*).

The water shrew has several locally-used synonyms including water ranny, otter shrew and labhallan (Scottish Gaelic).

2.2 Identification

Three species of shrew are found on mainland Britain – the common, pygmy and water shrew. These are red-toothed shrews, so-called because they (along with other members of the sub-family Soricinae) have red-tipped teeth due to the deposition of iron in the outer layer of enamel which may increase resistance to wear. The white-toothed shrews (Crocidurinae) lack this feature. The main identifying features of these three species are shown in Table 1.

Table 1 The key identifying features of the three species of shrew found on mainland Britain.

Features	Water shrew *Neomys fodiens*	Common shrew *Sorex araneus*	Pygmy shrew *Sorex minutus*
Head-body length	61-72mm	48-80mm	40-60mm
Head	Not particularly domed	Not particularly domed	Domed
Tail	Prominent keel of stiff silvery hairs under the tail. Used like a rudder during diving. Tail about as long as head-body.	A few stiff hairs under the tail. Tail 50-60% the length of the head-body.	A proportionally longer, thicker and more hairy tail than the common shrew. Tail 65-70% the length of the head-body.
Pelage colour	Black fur on back and silvery grey underside. Clear demarcation in adults. All-black individuals are not uncommon.	Three-tone coat colour. Dark brown fur on back, paler sides leading to a paler still underside.	Two-tone coat colour with darker back and paler underside. A paler colour than the common shrew.
Eyes/ears	Often has a white patch above the eye and white ear tufts.	No eye patches; occasionally with white ear tufts.	No eye patches or white ear tufts.
Weight	12-18g	5-14g	2.4-6.1g
Colour of teeth	Red-tipped	Red-tipped	Red-tipped
Number of unicuspid teeth (only 1 cusp per tooth)	4	5	5
Upper edge of first lower tooth	Smooth	Wavy	Wavy

Figure 1 Lower jaw of water shrew showing the red-tipped teeth.

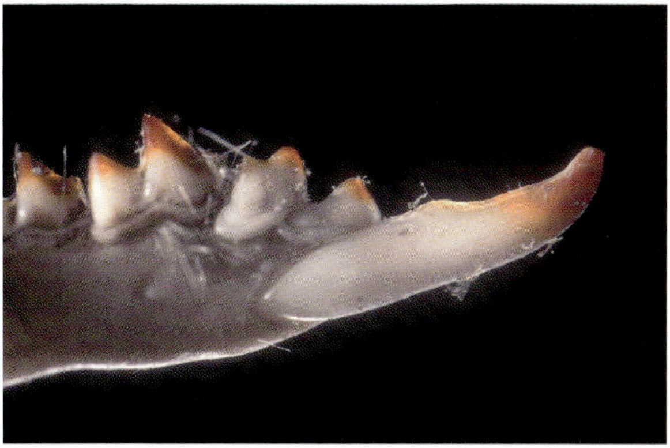

Figure 2 Comparison of the lower jaws of water shrews and common shrews.

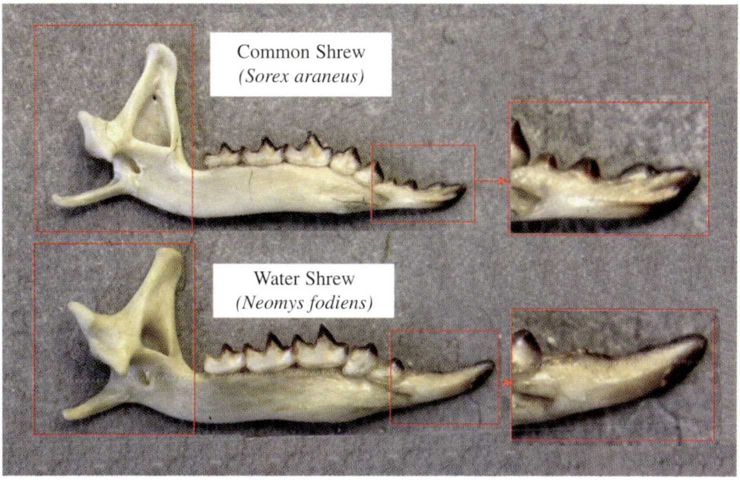

Difference in shapes of articular surfaces

Presence of 3 cusps on the incisor of the common shrew. Absence of cusps on the water shrew incisor

Figure 3 Water shrew eating earthworm. Note the clear demarcation between the dark back and white underside and note the white ear tufts.

Figure 4 Common shrew. Note the 3-tone coat colour.

Figure 5 Pygmy shrew. Note the 2-tone coat colour and the length of the tail in comparison to the body.

2.3 Breeding and young

Water shrews breed between April and September (Price, 1953; Churchfield, 1990). Gestation lasts between 19 and 21 days and females generally produce two litters a year. The average litter size is six, but can range from three to 15. Litter sizes decrease towards the end of the breeding season. Nests are usually made in burrows below ground and are formed from dried leaves and grass.

Figure 6 Young shrews in nest.

Young are fully weaned at about 27-28 days but in captivity have been shown to stay with the mother for up to 40 days after birth (Churchfield, 1979; Michalak, 1983). First litters of water shrews emerge from nests in May and once independent begin to disperse to establish themselves in their own home ranges.

2.4 Lifespan

Water shrews have an annual lifecycle and short life span (14-19 months), although they can live two years or so in captivity (Churchfield, 1984b, 1990). Shrews born in the summer pass the winter in an immature state, quickly growing to maturity the following spring when the breeding season commences. At the end of the breeding season, most adult water shrews die, leaving their young to carry the population through the winter. The populations thus follow a seasonal cycle with a peak in summer, a decline in autumn and low numbers throughout winter.

2.5 Activity

Water shrews do not hibernate, remaining active and foraging in water throughout the year. Activity above ground is reduced during winter when more time is spent in nests to reduce heat loss and help conserve energy. They are most active during darkness and show a peak in activity before dawn.

2.6 Home range

Water shrews are essentially solitary and generally occupy linear territories on the banks of watercourses. Territory size varies with season ranging from 77-173m^2 in winter to 101-373m^2 in summer (Lardet, 1988). They appear to spend only a short time (a few months) in one area before moving on (Churchfield, 1984a). They may recolonise the same area after an absence of several years. Water shrews can disperse widely and young shrews, in particular, can be found far from water (up to 3km).

2.7 Diving

All shrews can swim but the water shrew is the only British shrew that forages underwater. A layer of air, trapped between the skin and fur, insulates them from the cold water temperatures. To compensate for the buoyancy created by this insulating layer, they use all four limbs to paddle hard when diving. The surface area of each foot is increased by a fringe of stiff hairs, helping to propel them through the water. A keel of stiff hairs on the underside of the tail functions as a rudder. They may also use rocks and plants for anchorage, helping them to stay submerged. In the wild, dives usually last between three and ten seconds and are generally to depths of 30-200cm (Schloeth, 1980; Churchfield, 1998).

After diving, water shrews remove excess water from their coats by shaking, scratching with the hind feet and by squeezing through their narrow burrows.

Figure 7 Water shrew diving.

2.8 Food and foraging behaviour

Water shrews forage on land and in water throughout the year. To sustain their high metabolic rate they need to feed every two to three hours and must consume 50% of their body weight daily.

In freshwater habitats in southern Britain an average of 50% of the prey consumed by water shrews are aquatic invertebrates (Churchfield, 1985). The aquatic invertebrates eaten most frequently are water slaters (*Asellus* sp.), freshwater shrimps (*Gammarus* sp.), and caddis larvae (Trichoptera spp.). Aquatic prey is usually brought to the bank to be eaten on land. Water shrews can live entirely on terrestrial invertebrates when away from water and typical terrestrial prey includes beetles, spiders, centipedes, molluscs and earthworms. Millipedes are also eaten regularly by water shrews but not by common or pygmy shrews (Table 2).

Water shrews can take larger prey such as small fish, newts and frogs. These prey are immobilised by venom that is produced in the saliva from the submaxillary glands. The venom is neurotoxic, causing limb paralysis and blood and respiratory disorders in the prey. The toxin is non-lethal to humans but it can cause skin irritation and discomfort for several days.

Table 2 The major prey items found in the diets of water, common and pygmy shrews inhabiting watercress beds (after Churchfield, 1984a).

	Prey type	Water shrew	Common shrew	Pygmy shrew
Terrestrial prey	Adult Coleoptera (beetles)	•	•	•
	Hemiptera (bugs)	•	•	•
	Other insect larvae	•	•	•
	Myriapoda (centipedes)	•	•	•
	Myriapoda (millipedes)	•	•	•
	Isopoda (woodlice)	•	•	•
	Araneae and Opiliones (Spiders and harvestmen)	•	•	•
	Gastropoda (slugs and snails)	•	•	•
	Lumbricidae (earthworms)	•	•	
Aquatic prey	Trichoptera larvae (caddis larvae)	•		
	Other insect nymphs and larvae	•		
	Crustacea (incl. *Asellus* and *Gammarus* spp.)	•		
	Aquatic snails	•		
	Diptera larvae (fly larvae)	•		

2.9 Prey availability

Figure 8 Examples of the aquatic invertebrates commonly eaten by water shrews (A. Asellus, B. Gammarus, C. Trichoptera larvae and D. Trichoptera larvae cases). The colour of these invertebrates is an artefact of the photography.

The availability of prey is predicted to affect water shrew numbers and occurrence but few studies have investigated this. In a study of 41 sites along the River Medway and some of its tributaries in Kent, the presence of freshwater crustaceans was found to be one of the major factors determining the presence of water shrews (French *et al.*, 2001). At 27 other sites in Sussex (in the Ouse, Cuckmere and Pevensey Level catchments), water shrew occurrence was not associated with the presence of freshwater crustaceans but rather with a range of aquatic snails and Trichoptera larvae (Southgate, 2006). While the availability of aquatic prey may well influence the occurrence of water shrews at a site, their ability to feed on a diversity of both aquatic and terrestrial prey coupled with their nomadic tendencies probably enables them to survive even if watercourses dry up and they have to move to another area.

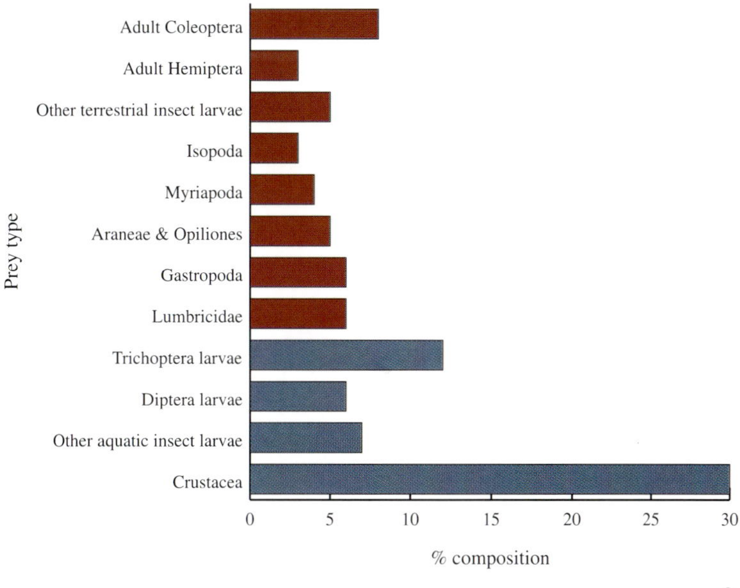

Figure 9 The percentage composition of major prey types in the diet of water shrews inhabiting watercress beds in southern England (After Churchfield, 1984a). Blue indicates aquatic prey; brown indicates terrestrial prey.

3 Water shrew distribution

3.1 World wide distribution

The water shrew is a Palaearctic species with a wide geographical distribution. It is found throughout northern and central Europe, Scandinavia, and eastwards across Russia into Siberia as far as Lake Baikal. Relict populations exist in parts of the Russian Far East, bordering the Sea of Japan. It is absent from most of Spain, Italy and Greece.

3.2 UK distribution

Water shrews are widely distributed across mainland Britain but are scarcer in northern Scotland (Churchfield, 1991; Carter & Churchfield, 2006). They are absent from Ireland and the Isle of Man but present on Anglesey, the Isle of Wight and a few of the Scottish islands (including Hoy, Raasay, Skye, Pabay, Mull, S. Shuna, Garvellachs, Islay, Kerrera and Arran).

Recent surveys suggest they are widespread on the UK mainland, especially in southern and central England (French *et al.*, 2001; Greenwood *et al.*, 2002; Carter & Churchfield, 2006; Southgate, 2006). Results from The Mammal Society's national Water Shrew Survey (2004-2005) (Carter & Churchfield, 2006) suggested that there was a greater occurrence of water shrews in the east of the country. This new finding needs further investigation but may be related to various factors including climate, altitude, topography and the type of habitat available (mostly lowland riparian).

Figure 10 Historical records of water shrews from 1993 to present day. Records collected during The Mammal Society's Water Shrew Survey (2004-2005) are not included

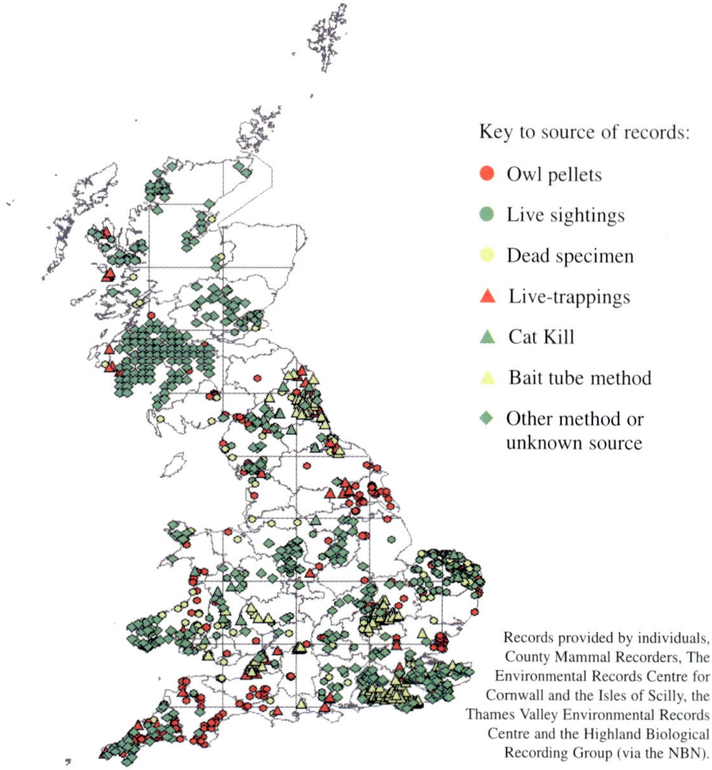

Figure 11 Water shrew distribution in Great Britain based on records collected during The Mammal Society's Water Shrew Survey (2004-2005).

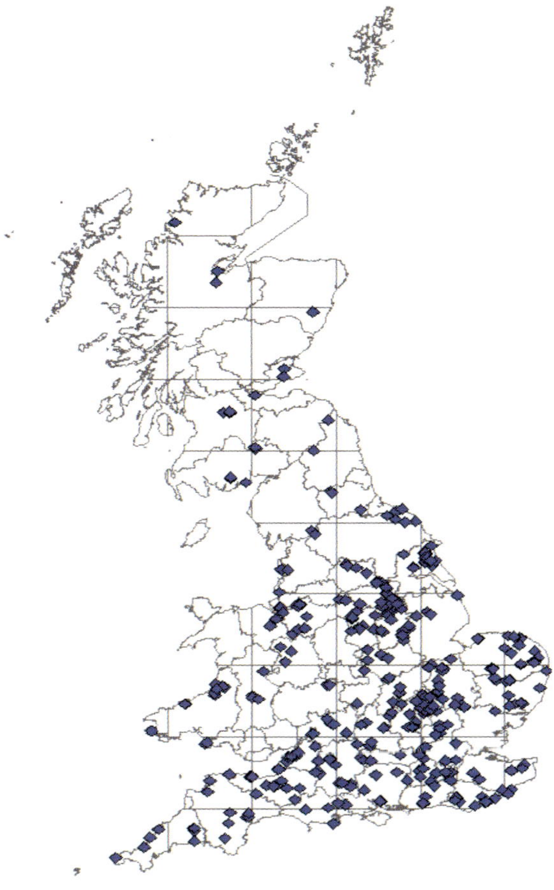

4 Water shrew habitat

Water shrews show great flexibility in their habitat occurrence and occupy many different habitats. Their tolerance of a range of habitat and environmental conditions suggests that they are a highly adaptable and versatile species. The following provides a brief overview of the habitats that water shrews are found in.

4.1 Habitat type

Water shrews are found along the banks of fast-flowing rivers and streams, by slow-flowing and static waters such as canals, ponds, lakes and ditches, and in fen, marsh and reed bed areas (Greenwood *et al.*, 2002; Carter & Churchfield, 2006; Southgate, 2006). They have even been found at brackish water and tidal sites (Carter & Churchfield, 2006; Southgate, 2006). Watercress beds have long been considered a favoured habitat of the water shrew (Churchfield, 1984b), although habitat modification (replacing grass banks with concrete), increased mechanisation and pesticide use in many commercial watercress beds have reduced their suitability for water shrews.

Table 3 The range of habitats in which water shrews were recorded during The Mammal Society's Water Shrew Survey 2004-2005 (Carter & Churchfield, 2006).

Habitat type	Total number of sites surveyed	Number of positive sites	% positive sites
River	410	65	15.6
Stream	666	114	17.2
Canal	78	24	30.8
Pond/Lake	420	77	18.3
Ditch	245	49	20.0
Bog	15	0	0
Fen/Marsh	35	9	25.7
Reed bed	41	7	17.1
Cress bed	2	0	0
Other	89	21	23.6

Figure 12 Typical water shrew habitat

4.2 Substrate type

The type of substrate found at an aquatic site is determined largely by the flow of the water. With minimal water flow, static and slow-flowing water bodies typically have a substrate of silt which supports *Asellus* spp. and several species of caddis fly larvae. The substrate of fast-flowing waters is typically rocks and stones and supports gammarids (freshwater shrimps) that are relatively intolerant of low oxygen levels. Water shrews can, therefore, be found foraging for prey in most substrates.

4.3 Water depth and width

Water shrews can be found in sites with water depths ranging from less than 0.25m to greater than 2m (Greenwood *et al.*, 2002; Carter & Churchfield, 2006). Foraging is probably restricted to the shallow water at the edge of a water body where currents are reduced and less energy is needed for diving and finding prey. The territory of a water shrew is linear along the banks of the watercourse (Lardet, 1988; Cantoni, 1993). Their occurrence is not restricted by width of water-course and they can be found in aquatic sites ranging from less than 1m to in excess of 10m wide (Carter & Churchfield, 2006).

4.4 Bank characteristics

Figure 13 Water shrew using steep, rock bank to gain access to water for foraging.

Water shrews are good climbers and very agile, capable of using banks with steep inclines (>45°) to enter and leave the water (see Figure 13). They are known to inhabit sites with bank heights ranging from less than 1m to greater than 2m (Carter & Churchfield, 2006). In areas prone to flooding, they may favour high banks in which to build their burrows, minimising the risks of water-logging if water levels rise (Greenwood *et al.*, 2002).

Earth banks provide a workable substrate in which water shrews can burrow, using their forefeet and teeth to loosen the soil, and their hind feet and mouth to remove soil and small stones (Churchfield, 1979). They also occupy sites with rock banks where gaps between the rocks may be used for shelter or nesting (Carter & Churchfield, 2006).

4.5 Aquatic and bankside vegetation

Water shrews are recorded most often in sites where aquatic vegetation is present but this is not essential as they are also found in sites where this vegetation is absent (Carter & Churchfield, 2006). The relationship between water shrews and terrestrial bankside vegetation is equally ambiguous. Dense herb cover was found to be important to water shrews in the national survey (Carter & Churchfield, 2006). Southgate (2006) found that reeds and grass/sedge tussocks were important to water shrews in east Sussex, while French *et al.* (2001) found that short grass had a negative impact on the species presence in the River Medway catchment (French *et al.*, 2001).

Water shrews are found at sites both with and without trees but do not favour areas with dense shrub and tree cover, as this shades out the ground cover on which they rely (Greenwood *et al*., 2002; Carter & Churchfield, 2006; Southgate, 2006).

4.6 Adjacent land use

Sites adjacent to arable land, cattle and sheep grazing, improved and unimproved grassland and broadleaved woodland all support water shrews (Carter & Churchfield, 2006; Southgate, 2006).

5 Population size

Water shrews are found at much lower population densities than either common or pygmy shrews. Maximum densities of three to five per hectare have been recorded in watercress beds (Churchfield, 1984b), while, depending on the habitat type, common shrews are found at densities of 17-69 per hectare (Shillito, 1960), and pygmy shrews at five to 30 per hectare (Churchfield & Brown, 1987). Densities vary according to site and season and their elusive nature and localised and sporadic distribution may have resulted in their densities being underestimated. Harris *et al*. (1995) estimated the British population to be 1,900,000 in 2004 but further studies are needed to assess the current population densities of water shrews in a range of habitats.

6 Legal status

Shrews are listed under Schedule 6 of the Wildlife and Countryside Act 1981. Accordingly, shrews may not be intentionally trapped without a licence from the relevant statutory body and, when trapping for other small mammals, precautions are necessary to minimise the chances of death or injury to shrews. *Appendix 1* gives details on obtaining a shrew licence.

7 Potential threats to water shrews

7.1 Habitat loss and modification

Several changes to habitats and habitat management practices could adversely affect water shrew populations:

- Reclamation of marsh and reed bed habitats for agriculture.
- Replacement of earth surface drains, in which water shrews are known to occur, by brick/concrete/plastic drainage pipes.
- Creation of concrete and other artificial banks on riversides.
- Replacement of grass banks with concrete in commercial watercress beds (a favoured habitat of water shrews) to allow motorised access.
- Burrow destruction during mechanised maintenance work along watercourses.
- Over-management of bankside vegetation leading to lack of shelter for water shrews moving along the bank.
- Over-grazing by domestic livestock removes the protective vegetative layer and poaching and trampling compacts the banks, making them unsuitable for burrowing.

- Changing water levels in river systems can result in flooding of banks and burrow systems of water shrews.
- Drainage causing seasonal drying of wetland and/or watercourses.

7.2 Pollution and pesticides

The water shrew is vulnerable to pollution and pesticides. Pollutants, such as heavy metals, can be ingested during grooming, while insecticides and molluscicides are readily accumulated by the invertebrates on which water shrews feed. Due to their high metabolic rate and high rate of food consumption, water shrews can quickly accumulate these toxins (Dodds-Smith *et al.*, 1992; Pankakoski *et al.*, 1994). Various biochemical factors such as pH, nitrate and phosphate levels and the availability of oxygen in a water body can affect the type and composition of prey and aquatic plants at a site. Results of The Mammal Society's national Water Shrew Survey showed that water shrews are more likely to be found in sites where the pH is in the range of 7.0-8.0, and where nitrate levels and Biochemical Oxygen Demand (BOD, a good indicator of organic pollution) are low (Carter & Churchfield, 2006). High BOD values indicate high rates of decomposition of organic matter and oxygen use by micro-organisms, reducing the oxygen available to other organisms in the water. There is some evidence that low Dissolved Oxygen and high water temperature are associated with absence of water shrews (Southgate, 2006), though it should be noted that the sites surveyed experienced a narrow range of water quality values. The pattern of distribution of water shrews in 96 sites in the Weald of S.E. England and their absence from sites with poor water quality (based on Environment Agency biological General Quality Assessment grades) suggested that pollution may be an important factor affecting these animals (Greenwood *et al.*, 2002). The biological GQA system is based on the macro-invertebrate communities of rivers and canals (see www.environment-agency.gov.uk for more details). However, more detailed analyses at local/regional and national levels have not confirmed an association between water shrew presence and quality of Environment Agency biological GQA-graded sites: water shrews were found in sites graded A-E (very good to poor) although not in an F-graded (very poor) site (Southgate, 2006; Carter, Churchfield & Barrow, unpubl. data). Again, sampling bias may contribute to these results: most sites surveyed were graded from A-C on the GQA scale.

7.3 Persecution and disturbance

With the possible exception of the occasional consumption of fish fry in commercial fish hatcheries and small fish, tadpoles, frogs and newts in garden ponds, water shrews do not conflict with man's interests, are not regarded as pests and are not directly subject to persecution. It is not known if their populations are affected by rodenticides used for brown rat control. Water shrews are most active at dawn and dusk: periods that do not coincide with peak human activity. They appear tolerant of low impact human activity and are commonly found in areas used by walkers, cyclists and anglers (Carter & Churchfield, 2006). They may be less tolerant of higher impact activities, although they have been recorded in gardens, near sports fields and even within a few hundred metres of a motorway, where human activity is more intense (Carter & Churchfield, 2006). While being relatively tolerant of human activity, it cannot be assumed that uncontrolled human activity will have no impact on water shrew populations.

7.4 Predators

Water shrews are predated upon by a variety of raptors but do not constitute a large proportion of the diet of any species. Between 1993 and 1997, 48,996 prey remains were retrieved from barn owl pellets collected from across Britain as part of The Mammal Society's National Owl Pellet Survey. Only 275 of these prey remains (0.56%) were identified as water shrews (Love *et al.*, 2000). Non-avian predators include pike, stoats, weasels, foxes and cats. Otters and mink may also be occasional predators of water shrews, although there is no documented evidence to support this. Despite many studies involving analyses of mink and otter scats, water shrew remains have rarely been found. The strong scent produced by all shrew species as part of their intra-specific communication and social organisation makes them relatively unpalatable to mammalian carnivores with their highly-developed sense of smell. So, although they may be killed by mammalian predators they are rarely eaten. Although water shrews do not appear to be heavily predated by any particular species, removal of just one or two individuals from their small, localised populations may have a significant effect on the dynamics of that population. It is possible that the impact of predation is underestimated: many water shrews may be killed by predators but, because they are not eaten, there is no evidence of them in scats or pellets.

8 Conservation and management advice

Although there has been no documented evidence of a population decline, studies have suggested that the water shrew has decreased in number and occurrence in locations where it was once abundant. The loss and degradation of riparian habitat and increases in pollution and pesticide use have been cited as possible causes for this speculative decline. Recent surveys have shown that water shrews are still widely distributed in aquatic habitats on mainland Britain but their populations are often localised and patchy.

Only 17.4% of survey sites nationally recorded the presence of water shrews, but on a regional scale in south-east England 42-82% of aquatic sites showed evidence of their presence (French *et al.*, 2001; Greenwood *et al.*, 2002; Carter & Churchfield, 2006; Southgate, 2006). So, although water shrews may have declined in recent years they are still widespread and may locally be relatively abundant. It should be stressed that nowhere over its geographic range is *N. fodiens* a common species and studies in Britain, continental Europe and Russia all agree that it is amongst the least abundant of all shrew species in the many habitats surveyed. Even in riparian habitats in Britain it is outnumbered by the common shrew and often also by the pygmy shrew. As a result of its relatively low population densities and its particular association with aquatic habitats it has probably been under-recorded, whether by trapping, owl pellet analysis, cat kills or sightings. With the design of new survey methods that are easy and economical to use, more information can be gathered about its population status.

Evidence concerning the role of habitat management in determining water shrew presence is equivocal. Although bank management (mowing to maintain short grass) is suggested to have an adverse effect on water shrew occurrence (French *et al.*, 2001) other studies have found these shrews to be tolerant of regular mowing and other bankside management (Churchfield, 1984b; Carter & Churchfield, 2006).

With such an ecologically flexible species it is difficult to make clear recommendations about habitat management, and the data on population densities currently available are not sufficient to gauge its conservation status accurately. However, some key features to be taken into account when developing habitat management strategies for conserving and promoting water shrews are as follows:

1. There is a positive association between water shrew presence and reed and grass/sedge tussocks, and with dense bankside ground cover.
2. There is a negative association between water shrew presence and with dense bankside tree and shrub cover and with short grass.
3. The presence of aquatic vegetation favours water shrews.
4. There is a positive association between water shrew presence and the abundance of aquatic crustaceans and the presence of aquatic snails and caddis larvae.
5. There is no association between water shrew presence and Environment Agency biological GQA grades, phosphate, orthophosphate or ammonia, but there is a negative association between water shrew presence and low dissolved oxygen, high water temperature, high BOD and high nitrate levels. There is a positive association with water of pH 7.0-8.0.

The wide geographical distribution and broad habitat range of water shrews are encouraging but should not induce complacency. Poor water quality and loss or degradation of riparian habitat may affect water shrew occurrence on a local scale and there is a possibility that the spread of predatory American mink may adversely affect their distribution. To assess their conservation needs fully it is necessary to determine and monitor the population sizes of water shrews in Britain.

We would encourage sensitive management of riparian sites to promote water shrew populations. This should include the maintenance of bankside ground cover, the management of bankside trees and shrubs to reduce shading and encourage growth of ground cover, and the maintenance of high dissolved oxygen, low BOD and nitrate levels and pH 7.0-8.0 in watercourses, which will support growth of invertebrate prey populations.

9 Water shrew surveys

9.1 Why survey?

Surveys for water shrews are necessary to determine the distribution and habitat occurrence of water shrews at a national, regional and local level. The data from these surveys provide essential information with which future changes to water shrew populations can be compared.

9.2 Field signs to look for

Water shrews are occasionally sighted when diving but, when viewed at a distance swimming on the water surface, they can be confused with common shrews and bank voles that are also accomplished swimmers.

The small size of water shrews means that their field signs are discreet and easily overlooked. However, careful searching of a site can reveal some key field signs:

Footprints

Like humans, water shrews have a simple plantigrade foot, running with the soles of their hind feet placed flat on the ground. Each foot possesses five toes or digits, each terminating in a simple claw.

Careful examination of the soft margin of a watercourse may reveal the footprints of several mammal species. These can be easily distorted and overlaid by the footprints of other individuals, making accurate species identification using this method difficult.

Burrows

Water shrew burrows are usually sited well above water level to prevent flooding. They either dig their own burrows or use the disused burrows of species such as bank voles and wood mice. The burrow entrances of water shrews are small (about 2cm in diameter) and the vegetation around the entrance remains intact unlike those of small rodents which are scuffed bare of vegetation.

Figure 14 Water shrew burrow. Note size in relation to the clover.

Feeding signs

Caches of prey remains may be found at feeding sites along the water's edge. Common items that can be found include piles of broken mollusc shells, the larval cases of caddis flies and even small twigs and stones mistaken as prey during foraging dives. Feeding sites are often small areas of earth or gravel or flat stones located under an overhang or in a secluded crevice.

Figure 15 Larval case of caddis fly.

Figure 16 Remains of aquatic snails.

Piles of snail shells may also be left by brown rats or wood mice, so to confirm the presence of water shrews, other prey remains and key field signs should be searched for.

Scats (faeces)

Water shrew scats are distinctive (see Table 4) but are difficult to find due to their small size. They can be found singly or in piles on rocks along the water body, near feeding sites, in surface run ways or close to burrow entrances.

Vocalisation

Water shrews, like other British shrews, produce loud, high-pitched but audible squeaks in a rapid, continuous, repetitive sequence, plus a characteristic "churr-churr" at a lower frequency. These are used as threats or warning signals to other shrews. Vocalisation is most common during the breeding season and when young are dispersing to find new territories. Differentiating between the calls of shrew species is not easy but those of water shrews are lower pitched and can be reliably distinguished with practice.

9.3 Methods for surveying water shrews

With field signs difficult to find and distinguish, and sightings rare, it may be necessary to use one of the available survey methods to determine the presence of water shrews. The best method for each individual survey should be chosen based on the type of data required (*e.g.* abundance or distribution data), survey effort, surveyor experience and cost. Given the widespread distribution of water shrews it is worth surveying a range of different aquatic habitat types.

a) Live-trapping

Before undertaking any live-trapping work for shrews, all surveyors should be experienced in methods of live-trapping and in handling small mammals, and should hold a shrew licence (see *Appendix 1* – Shrew licences). The surveyor needs to be able to distinguish water shrews from any other species which may be also be trapped.

Trap types

Box traps such as Longworth traps and Sherman traps can be used. Both of these types of trap are effective for catching water shrews, although Longworth traps tend to be used most widely in the UK. Sherman traps do not have a separate nest box so bedding and food can obstruct the trap mechanism, preventing the door from closing and allowing captives to escape.

Bait

Shrews cannot survive for more than about three hours without food and so suitable food is needed to keep them alive in traps. Casters (blowfly pupae *Calliphora* sp) make good bait and survival-food for shrews and a handful will keep a shrew going for several hours. Casters should be frozen prior to use to prevent them hatching into flies during the survey period. Dried mealworms and minced beef are suitable alternatives but the latter will deteriorate quickly in warm weather. A small piece of apple or carrot should be provided as a source of moisture for trapped mammals.

Bedding

The small size of shrews means that they lose heat rapidly when wet and are at risk of dying. Using a suitable bedding will greatly reduce the mortality of shrews in traps. In winter, and in wet weather, non-absorbent cotton wool is a better insulator than hay or straw, which can be used as bedding in the summer.

Where to place the traps

Place traps in vegetation as close to the edge of the chosen water body as possible (within 2-3m). Traps should be secured firmly so that they do not fall into the water when a small mammal is caught. To prevent traps being flooded by rising water levels make sure they are placed well above water level. In summer, ensure the traps are shaded by vegetation to minimise mortality from overheating.

Checking traps

Trap mortality is more of a problem for shrews than for other small mammals as a result of their small size and high metabolic rate. They quickly die of starvation and readily succumb to excesses of cold and heat (particularly in metal live-traps). It is advisable to check traps a minimum of four times in a 24-hour period (dawn, midday, late afternoon and evening). In extreme weather (*i.e.* very cold or very hot), and during the breeding season, traps should be checked more often.

Handling the capture

All captures should be handled as little as possible to minimise stress. Water shrews can bite and a toxin in their saliva can cause reddening and irritation to the skin for several days (see *Appendix 3* – Health and safety information).

Data that can be collected using live traps

a) Water shrew presence - The presence of a water shrew in a trap indicates their occurrence at a site but the lack of a water shrew in a trap is not indicative of their absence. They may be present but just unrecorded during the survey.

b) Abundance data - Capture-mark-recapture (CMR) allows the size of a water shrew population in a single location to be determined. Fur-clipping is a suitable short-term marking technique. Different patterns of fur clipping of the guard hairs on the back and sides can be used to identify individual water shrews (see Gurnell & Flowerdew, 2006).

The information provided here is only a brief summary of how to live trap for water shrews. A fully comprehensive guide to small mammal trapping can be found in Gurnell & Flowerdew (2006).

b) Bait tube method

Bait tubes are a cheap and effective method for determining the presence of water shrews at a site (Churchfield *et al.*, 2000). Small mammals (including water shrews) are encouraged to enter a piece of plastic tubing supplied with bait. They are free to enter and leave the tube at will. As they linger to explore and eat the bait they deposit scats. The form of the scat and its contents allow different species of small mammal to be identified. Accurate identification of water shrew scats usually relies on the presence of fragments of freshwater invertebrates in the scats and so the technique is most effective in aquatic habitats.

This method has been used successfully in several local, regional and nationwide surveys including The Mammal Society's Water Shrew Survey (2004-2005) (French *et al*., 2001; Greenwood *et al*., 2002; Carter & Churchfield, 2006; Southgate, 2006). It is reliant on the surveyor being able to accurately identify water shrew scats.

Bait tubes

Bait tubes are made from 4cm diameter plastic waste pipe which can be purchased from most DIY shops. White pipe is recommended as tubes of this colour are easier to find amongst vegetation and recover at the end of the survey. The optimum tube length recommended is 20cm and the cut edges should be sand-papered to minimise abrasion. One end of each tube should be covered with a small piece of muslin or nylon (old tights are good for this), and secured in place with an elastic band. These covers prevent bait falling out of the tubes when they are placed at a site. It is common for small mammals to gnaw through these covers to gain access to the bait.

Bait

Pre-frozen casters or dried mealworms can be used in bait tubes and 20-30 should be placed in each tube. No bedding is required as small mammals are not trapped and can enter and leave the tubes at will.

Where to place the tubes

Bait tubes should be placed horizontally within 2-3m of the water's edge, their entrances flush to the ground and hidden from sight in the vegetation. Tubes placed in open, exposed places rather than amongst bankside vegetation are less likely to be visited. The tubes can be secured in place with a loop of wire and their position marked with a piece of brightly coloured wool or twine tied to nearby vegetation (this should be removed and disposed of appropriately on completion of the survey). All tubes should be placed high enough above the water level to prevent flooding. In bog and fen sites tubes should be placed in dry areas, such as on grass tussocks.

It is recommended that no fewer than four bait tubes are used at a single site. This number has been shown to be sufficient for detecting the presence of water shrews at a site (Carter & Churchfield, 2006). However, water shrews can be highly localised in occurrence and the more tubes used, the greater the likelihood of success: use as many tubes as possible to maximise the sampling area. As a guide, the nationwide survey (Carter & Churchfield, 2006) adopted a protocol of placing bait tubes at 10m intervals at a site. The tubes can be left in place for two weeks without checking or re-baiting. Following collection, the tubes should be air-dried and the contents carefully scraped out and placed in containers (film canisters are useful) labelled with the date and details of the site. Scats will keep indefinitely when dry but will degrade and decompose rapidly if wet.

Figure 17 Bait tube in bank side vegetation.

Scat identification

Drying scats aids their identification. Shrew scats should be separated from any rodent scats in the sample. Shrew scats are identifiable by their granular consistency and the ease with which they crumble when pressed gently. By comparison, rodent scats are extremely tough and fibrous, and are much more resistant to pressure (see Table 4).

Table 4 Key features to separate rodent and shrew scats

Identifying features	Rodent scats (excluding water vole and brown rat)	Terrestrial shrew scats (common and pygmy shrews)	Water shrew scats
Colour	Black/brown in colour when both wet and dry.	Black/grey in colour when both wet and dry.	Black/grey when wet. Pale grey/ silvery when dry.
Outer surface	Smooth	Uneven and irregular	Uneven and irregular
Shape	Cylindrical & pointed at one end.	Elongated with a sharp point at one end.	Oval and rather blunt-ended.
Size	3-8mm in length Approx. 1-2mm wide.	2-5mm in length Approx. 1-2mm wide.	3-10mm in length Approx. 2-3mm wide.
Consistency	Very hard when dry – do not crumble. Fibrous texture when broken apart.	Crumble easily when dry - granular texture due to many, fine fragments of terrestrial arthropod remains.	Crumble easily when dry - granular texture due to many fragments of aquatic and terrestrial prey remains.

If the scats are whole, those of water shrew and common/pygmy shrew can often be distinguished with the naked eye or with a hand lens on the basis of the characteristics in Table 4 (see Figure 18). Water shrew scats frequently possess relatively large shards of grey-white chitin from aquatic crustaceans and/or terrestrial millipedes. These are clearly visible with the aid of a hand lens as they protrude from the surface of the scat.

Figure 18 Common shrew scats (left) and water shrew scats (right) (x 8 magnification). Note difference in size and colour between species.

It is necessary, however, to confirm their identity on the basis of the prey they have eaten and this should be done using a binocular dissecting microscope. Individual scats should be broken apart in a drop of water in a Petri dish. Whole prey items will not be found in scats so it is necessary to identify fragments of the aquatic and terrestrial invertebrates most commonly consumed by shrews. Examples of some of the prey items eaten by water shrews can be found in *Appendix 2*. The presence of aquatic crustaceans (*Gammarus* or *Asellus*) and caddis larvae and terrestrial millipedes (Diplopoda) in the scats are the best diagnostic indicators of water shrews since these are eaten in large quantities in most aquatic habitats and they leave readily-identifiable fragments. Identifying water shrew scats from their prey fragments will take practice and is helped by collecting and examining reference samples of common invertebrates from the study site. These can be preserved in 70% alcohol for continued reference.

Data that can be collected using bait tubes

a) Water shrew presence - The presence of water shrews can be ascertained through the identification of scats. This method can be used to determine the distribution of the species in an area. While the occurrence of scats confirms the presence of water shrews, their absence from bait tubes does not necessarily mean that they are not present in the study area: they may not have encountered the bait tubes during the sampling period.

b) Abundance of water shrews - Population densities of water shrews at a particular site cannot be estimated using bait tubes on the basis of the protocol outlined above because there is no way of knowing how many different individuals have visited the bait tubes if they are placed close together at a site. A water shrew may travel up to 200m in a day along the bank of a watercourse so if bait tubes are placed only 10m apart, it is highly likely that the same water shrew has visited each one. In order to increase the chances of recording the presence of different individuals, sampling sites must be distributed at least 200m apart. If several bait tubes are placed at a single location (at, say, 10m intervals) it is safest to assume that all scats deposited are from the same shrew. Using a protocol modified in this way a conservative estimate of abundance of water shrews within a target study area can be made.

c) Hair tubes

Hair tubes comprise pieces of plastic tubing of 35mm diameter cut to lengths of approximately 120mm. Each tube is then cut in half lengthways and double-sided sticky tape (*e.g.* carpet tape) is applied to the interior of the top half. The two halves are taped together and the tube placed on the ground in the chosen site. Hairs of small mammals are trapped on the sticky tape as they pass through the tube. The outer guard hairs adhere to the sticky tape and can be removed and identified according to their colour, length, cross section and scale pattern. Details of how to prepare and examine hair samples and the key identifying features to look for can be found in Teerink (1991) and Ducommun *et al.* (1994). The hairs of common, pygmy and water shrews are remarkably similar and those of water shrews can only be distinguished reliably by examining the cross sections of hairs. The guard hairs of water shrews have a more pronounced H-shape with many ridges in the lateral grooves (Hutterer & Hürter, 1981). Preparing and examining cross sections of hairs is laborious. Recent work has found that specific measurements can be taken from the guard hairs of shrews, enabling them to be identified to species level with 85% accuracy (Pocock & Jennings, in press). An index of abundance of water shrews can be estimated if hair tubes are placed at suitable distances apart (see above).

The analysis of hair samples by any of these methods is time-consuming and requires practice. Current research is investigating the use of DNA analysis to identify hair samples. Although more costly than conventional techniques this should be a rapid method for identifying water shrews and other mammals from hair samples.

d) Owl pellets

The distinctive jaws of the water shrew (see Table 1; Figure 1; Figure 2) allow them to be distinguished relatively easily from those of other shrews and rodents. However, owls have large foraging ranges and the precise location of the caught shrew will not be known. Nonetheless, it will indicate the presence of water shrews in an area. For more details on identifying water shrews from owl pellets see Yalden (2003).

9.4 Recording habitat information

Whichever survey method is employed, habitat information should be collected at each site. This will help to build a picture of the habitat occurrence of water shrews within the survey area and the habitat characteristics that are associated with their presence. This information may be used to predict their occurrence in other similar sites, and will help in developing habitat management strategies to conserve and encourage populations of water shrews. The information can be recorded by filling in a simple pro-forma that summarises the habitat features at the survey site. Although the characteristics of a site will vary along its length, *e.g.* bank height, the predominant measurement or feature should be recorded. A typical recording form should include the following:

- Site details - Record the Ordnance Survey grid reference for the site and name each site after the watercourse or a distinctive local landmark.
- Habitat type - The habitat that best describes the watercourse or water-body, *e.g.* river, stream, pond, marsh, fen, ditch, canal, reed bed, watercress bed.
- Water depth - Estimate the depth (m) of the watercourse/water-body. This can be done from the shore with no need to enter the water.
- Width of water-body - Estimate the average width (m) of the water-body at the point that is being surveyed.
- Current - Assess the speed of the watercourse. Is it static (0m/s, no perceptible flow), slow-flowing (0.1-0.5m/s) or fast (>0.5m/s)?
- Substrate type - Note the predominant substrate type. Is it mostly rocks and stones (particles >20mm diameter), gravel (particles 2-20mm) or silt (particles <2mm)?
- Bank type - Record if the bank is made of earth or rocks.
- Bank incline - Record if the incline of the bank is at an angle of <45° (shallow) or >45° (steep) to the watercourse.
- Bank height - Estimate the average height of the bank (m).
- Bankside management - Record the frequency with which bankside vegetation is managed, *i.e.* never, occasionally or frequently. Record the type of management if known.
- Aquatic vegetation – Record the presence and relative amount of plant cover within the water-body in these three categories: a) submerged plants (those with growing parts mainly below the water surface, *e.g.* water buttercup, *Ranunculus*; water milfoil, *Myriophyllum*), b) emergent plants (with most of the vegetative portion growing above the water surface, *e.g.* reeds, *Phragmites*; rushes, *Juncus*), or c) floating (*e.g.* duckweed, *Lemna*; water lilies, *Nymphaea*). Aquatic vegetation at a site can be recorded simply as absent, present or dense or the commonly used DAFORN scale (Kent & Coker, 1992) based on the estimated percentage cover can be used, as follows:
    - Dominant 81-100%
    - Abundant 61-80%
    - Frequent 41-60%
    - Occasional 21-40%
    - Rare 1-20%
    - None 0%
- Terrestrial vegetation – Record the level of cover of bankside trees, shrubs, herbs/forbs and grasses using the scales above.

- Human activity – Record the type and level of human activity. Is the site used for walking, angling, boating or other pursuits? Is the use of the site light or heavy?

- Adjacent land use – Assess the type of land use adjacent to your site, *e.g.* urban, arable, broadleaved woodland, mixed woodland.

- Aquatic invertebrates – Determine the types and abundance of aquatic invertebrates using a long-handled pond-net. This can be done by 'kick sampling'(Mason, 2002) in which the sediment is stirred up with the foot while holding the net downstream to catch any dislodged invertebrates or by sweeping the net through the water and submerged vegetation. If identification is possible, record the abundance of *Asellus, Gammarus*, caddis larvae and molluscs – favoured prey items of water shrews. Abundance can be assessed as absent, present or dense, or it can be assessed more quantitatively by sampling with a pond net for a set time (say 2 minutes per sample). This will provide useful quantitative data for making comparisons between sites.

10 Acknowledgements

The authors would like to thank the Environment Agency's Science Programme for funding the production of *The Water Shrew Handbook* and for helping to fund The Mammal Society's Water Shrew Survey (2004-2005). We also thank Andrew Heaton and Henry Brown from the Environment Agency for their assistance during the survey and the production of this book.

All the data from The Mammal Society's Water Shrew Survey (2004-2005) was collected by volunteers to whom we are most grateful. We are also grateful to all the County Mammal Recorders who provided us with historical water shrew records. We are also grateful to the following people for their help at various times during the survey and the production of this book; Rachel Barrow, Gayle Dower, Sarah Gardner, Simon Poulton, Georgette Shearer, Fran Southgate, Phil Wheeler, Alison Tutt and Derek Yalden.

Finally, we would like to thank all the people who generously donated the photographs featured in this book; Yves Adams, Dr Steve Furness, David Hosking, Alasdair Love, Martin Smith and Rollin Verlinde.

11 References

Cantoni, D. (1993). Social and spatial organisation of free-ranging shrews *Sorex coronatus* and *Neomys fodiens* (Insectivora, Mammalia). *Animal Behaviour* 45: 975-995.

Carter, P.K.L. & Churchfield, S. (2006). Distribution and habitat occurrence of water shrews in Great Britain. Environment Agency Science Report SC010073/SR, The Environment Agency, Bristol and The Mammal Society Research Report No. 7, The Mammal Society, London (available to download at www.mammal.org.uk).

Churchfield, S. (1979). Studies on the ecology and behaviour of British shrews. PhD thesis, University of London.

Churchfield, S. (1984a). Dietary separation in three species of shrew inhabiting watercress beds. *Journal of Zoology* 204: 211-228.

Churchfield, S. (1984b). An investigation of the population ecology of syntopic shrews inhabiting watercress beds. *Journal of Zoology* 204: 229-240.

Churchfield, S. (1985). The feeding ecology of the European water shrew. *Mammal Review* 15: 13-21.

Churchfield, S. (1990). The Natural History of Shrews. Christopher Helm/A&C Black, London.

Churchfield, S. (1991). Family Soricidae (Shrews). In: The Handbook of British Mammals (eds. Corbet, G. B. & Harris, S.). 3rd Edition, Blackwell Scientific Publication. pp 49-75.

Churchfield, S. (1997). Species action plan for England and Wales. Water shrew – *Neomys fodiens*. National Rivers Authority/Environment Agency report.

Churchfield, S. (1998). Habitat use by water shrews, the smallest of amphibious mammals. *Symposium of the Zoological Society of London* 71: 49-68.

Churchfield, S., Barber, J. & Quinn, C. (2000). A new survey method for Water Shrews (*Neomys fodiens*) using baited tubes. *Mammal Review* 30: 249-254.

Dodds-Smith, M.E., Johnson, M.S. & Thompson, D.J. (1992). Trace metal accumulation by the shrew *Sorex araneus*. I. Total-body burden, growth and mortality. *Ecotoxicology and Environmental Safety* 24: 102-117.

Ducommun, M-A., Jeanmaire-Besançon, F. & Vogel, P. (1994). Shield morphology of curly overhair in 22 genera of Soricidae (Insectivora, Mammalia). *Revue Suisse de Zoologie* 101: 623-643.

Ellenbroek, F.J.M. (1980). Interspecific competition in the shrews *Sorex araneus* and *Sorex minutus* (Soricidae, Insectivora): a population study of the Irish pygmy shrew. *Journal of Zoology*, London 192: 119-136.

Environmental Records Centre for Cornwall and the Isles of Scilly (to XII/2005) *Erecords computer database*. Cornwall Wildlife Trust. Unpublished.

French, B.I., Mesquita, F. & Griffiths, H.I. (2001). Habitat variables affecting the occurrence of *Neomys fodiens* (Mammalia, Insectivora) in Kent, UK. *Folia Zoologica* 50: 99-105.

Greenwood, A., Churchfield, S. & Hickey, C. (2002). Geographical distribution and habitat occurrence of the water shrew (*Neomys fodiens*) in the Weald of South-East England. *Mammal Review* 32: 40-50.

Gurnell, J. & Flowerdew, J.R. (2006). Live Trapping Small Mammals: A Practical Guide 4th edition. The Mammal Society, London.

Harris, S., Morris, P., Wray, S. & Yalden, D. (1995). A Review of British Mammals: Population Estimates and Conservation Status of British Mammals other than Cetaceans. Joint Nature Conservation Committee, Peterborough.

Hutterer, R. & Hürter, T. (1981). Adaptive Haarstrukturen bei Wasserspitzmausen (Insectivora, Soricinae). *Zeitschrift für Säugetierkunde* 46: 1-11.

Kent, M. & Coker, P. (1992). Vegetation description and analysis: a practical approach. John Wiley, Chichester.

Lardet, J-P. (1988). Spatial behaviour and activity patterns of the water shrew *Neomys fodiens* in the field. *Acta Theriologica* 33: 293-303.

Love, R.A., Webbon, C.C., Glue, D.E. & Harris, S. (2000). Changes in the food of British Barn Owls (*Tyto alba*) between 1974 and 1997. *Mammal Review* 30: 107-129.

Mason, C.F. (2002). Biology of freshwater pollution. 4th Edition. Pearson Education, Harlow.

Michalak, I. (1983). Reproduction, maternal and social behaviour of the European water shrew under laboratory conditions. *Acta Theriologica* 28: 3-24.

Pankakoski, E., Koivisto, I., Hyvarinen, H. & Terhivuo, J. (1994). Shrews as indicators of heavy metal pollution. In: Advances in the Biology of Shrews (eds. J.F. Merritt, G.L. Kirkland & R.K. Rose), pp. 137-149. Carnegie Museum of Natural History Special Publication 18.

Pocock, M. J. O. & Jennings, N. (2006). Use of hair tubes to survey for shrews: new methods for identification and quantification of abundance. *Mammal Review* 36.

Price, M. (1953). The reproductive cycle of the water shrew, *Neomys fodiens* bicolor Shaw. *Proceedings of the Zoological Society of London* 123: 599-621.

Pucek, M. (1959). The effect of the venom of the European water shrew (*Neomys fodiens fodiens* Pennant) on certain experimental animals. *Acta Theriologica* 3: 93-103

Schloeth, R. (1980). Freilandbeobachtungen an der Wasserspitzmaus, Neomys fodiens (Pennant, 1771) im Schweizerischen Nationalpark. *Revue Suisse de Zoologie* 87: 937-939.

Shillito, J. F. (1960). The general ecology of the common shrew, *Sorex araneus L.* PhD thesis, University of Exeter.

Southgate, F. (2006). An analysis of the impact of wetland habitat and water quality on the distribution of the northern water shrew (*Neomys fodiens*) in Sussex. Thesis submitted for M. Phil., University of Brighton.

Teerink, B. J. (1991). Hair of west European mammals: Atlas and identification key. Cambridge University Press, Cambridge (*out of print*).

Vogel, P., Bodmer, C., Spreng, M. & Aeschimann, J. (1998). Diving capacity and foraging behaviour of the water shrew (*Neomys fodiens*). *Symposium of the Zoological Society of London* 71: 31-47.

Wilson, D.E. & Reeder, D.M. (eds) (2005). Mammal Species of the World, a taxonomomic and geographic reference. Vol. 1. 3rd Edition. John Hopkins Univ. Press, Baltimore.

Yalden (2003). Analysis of Owl Pellets. The Mammal Society, London.

12 Appendices

Appendix 1 - Shrew licences

You are required to hold a shrew licence if you are going to carry out live-trapping for shrews. Members of The Mammal Society can apply to be covered under The Mammal Society's block shrew licence. If you intend to mark or kill individuals, or you are not a member of The Mammal Society, you need to apply for an individual licence from the relevant statutory body (see below). It is usual to be asked for two referees who can vouch for your competency in live trapping and any other techniques that may be used. Licences are free.

To be covered by The Mammal Society's block licence you will need to:

- Be a member of The Mammal Society.
- Provide two written references from supervisors/colleagues, stating your competency in small mammal trapping.
- Have attended a course that covered small mammal trapping (please state course dates).
- Apply in writing to The Mammal Society Office stating the above and also stating in which country (i.e. England, Scotland or Wales) you will be trapping.

Statutory Nature Conservation Organisations:
English Nature/Natural England
Northminster House, Peterborough, PE1 1UA
T: 01733 455 000 W: www.english-nature.org.uk

Scottish Natural Heritage
2/5 Anderson Place, Edinburgh, EH6 5NP
T: 0131 447 4784 W: www.snh.org.uk

Countryside Council for Wales
Maes-y-Ffnnon, Penrhosgarnedd, Bangor, Gwynedd, LL57 2DW
T: 0845 130 6229 W: www.ccw.gov.uk

Appendix 2 - Examples of prey remains that can be found in water shrew scats

The following photographs show various body parts (*e.g.* limbs, antennae) of the prey most commonly consumed by water shrews. These body parts have been removed from whole specimens to show the main diagnostic features of each species. In scats, the body parts will be much fragmented. These photographs should be used to familiarise yourself with the key identifying features of the prey which will help when examining the fragments found in scats. It is suggested that they are used in conjunction with your own reference collection of prey invertebrates preserved in 70% alcohol.

All the fragments of prey in the following images have been photographed in a droplet of water and the accompanying descriptions apply to their appearance in water; dried fragments will have a slightly different appearance especially with regard to colour.

Water slaters *(Asellus)*

- Dominant prey item in the water shrew diet throughout the year; rarely, if ever, eaten by common or pygmy shrews.
- Legs are segmented with a few long, fine, spines protruding from each segment. The legs can be easily confused with those of terrestrial woodlice although the final segment before the terminal claw is longer in *Asellus.*
- The antennae are long with numerous, small segments.
- Carapace is transparent, thin and flexible (try bending it with a needle or fine seeker). By contrast, woodlouse carapace is thicker, heavily chitinised and relatively inflexible.

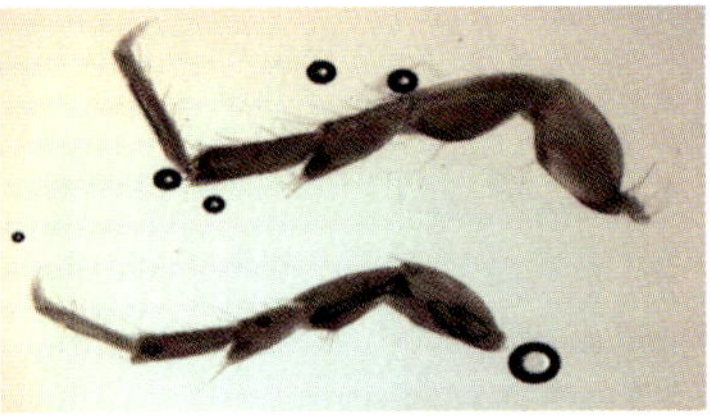

Asellus legs (x12 magnification)

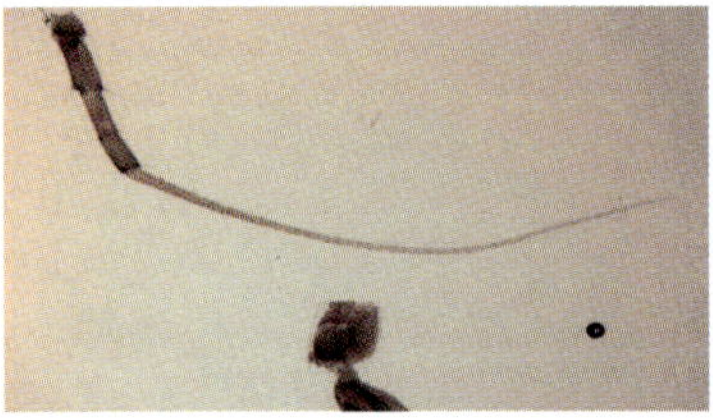

Asellus antenna (x12 magnification)

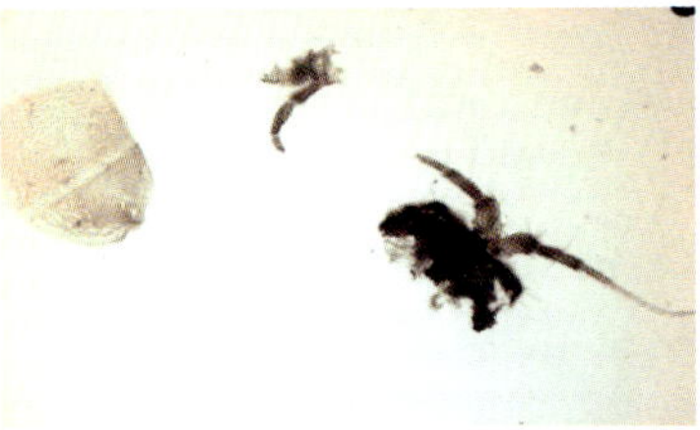

Asellus carapace (left) and head (right)

Freshwater shrimp *(Gammarus)*

- Abundant in the water shrew diet in summer and autumn; rarely, if ever, eaten by common or pygmy shrews.
- Many legs of differing shapes – all legs and appendages have dense fringes of long hair.
- Pale and translucent carapace.
- Long, well-segmented antennae.

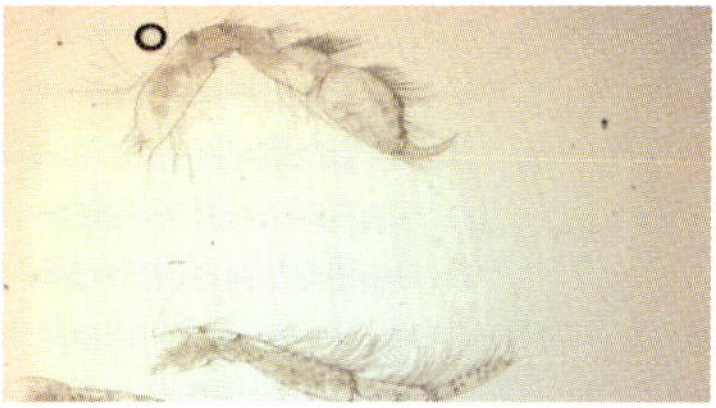

Gammarus legs (x6 magnification)

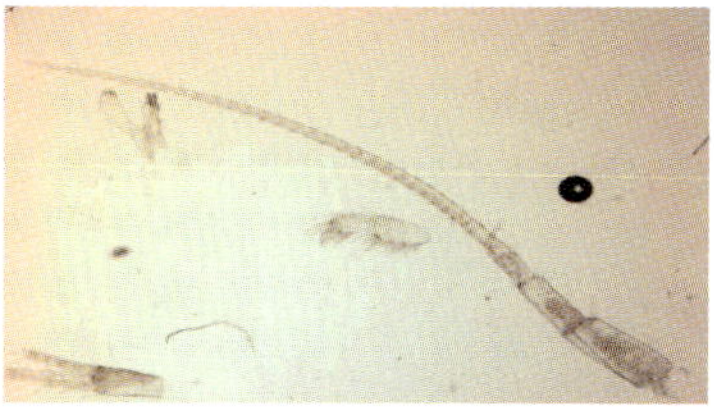

Gammarus antenna (x10 magnification)

Caddis larvae *(Trichoptera sp.)*

- Consumed by water shrews throughout the year; rarely, if ever, consumed by common or pygmy shrews.
- Legs are heavily chitinised and brown in colour. Long, fine, black hairs protrude from the legs.
- The claw at the end of each leg is highly curved.

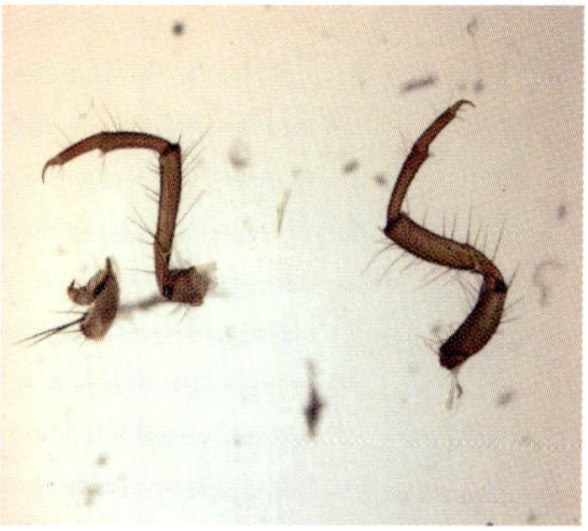

Caddis larva legs (x12 magnification)

Mayfly nymphs *(Ephemeroptera)*

- Occasionally consumed by water shrews; rarely, if ever, consumed by common or pygmy shrews.
- Legs are long and thin with a large, hooked claw.
- Tail comprised of three, long, many-segmented appendages.

Mayfly leg (x10 magnification)

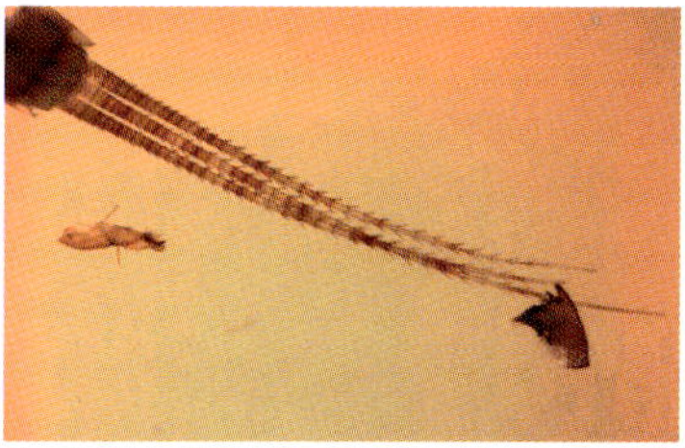

Mayfly tail (x10 magnification)

Millipedes *(Diplopoda)*

- Only consumed by water shrews; not eaten by common or pygmy shrews.
- Legs and antennae are heavily chitinised, thickened and often pigmented.
- Legs are curved with a well defined claw. Usually many of these legs are found in a scat.
- Antennae are short and thick with few segments.

- Carapace is thick, translucent and striated.
- In dried scats the carapace is distinct – thick, white and striated.

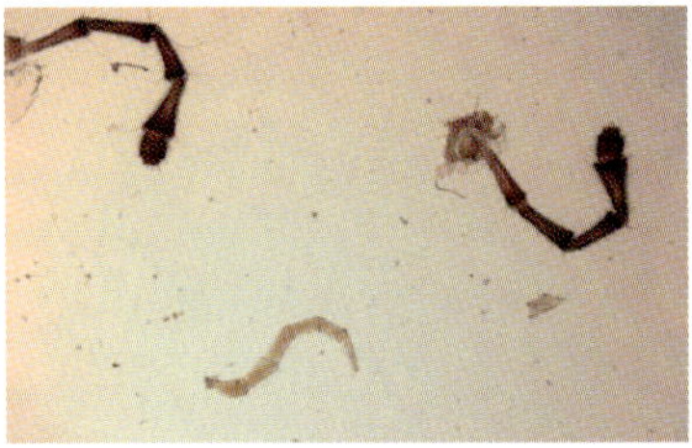

Millipede antennae (top left and right)
Leg (bottom) (x10 magnification)

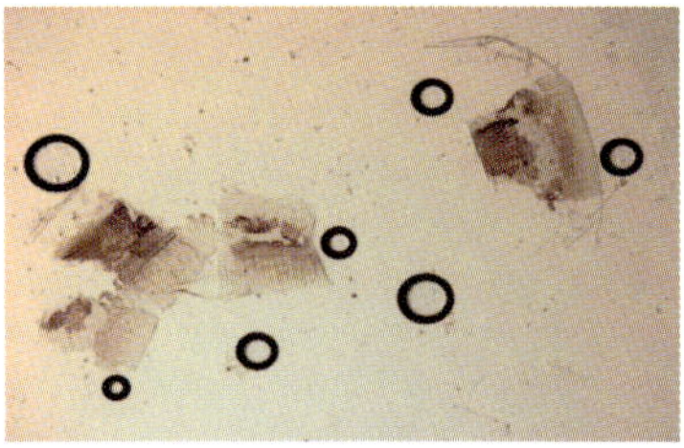

Millipede carapace (x10 magnification)

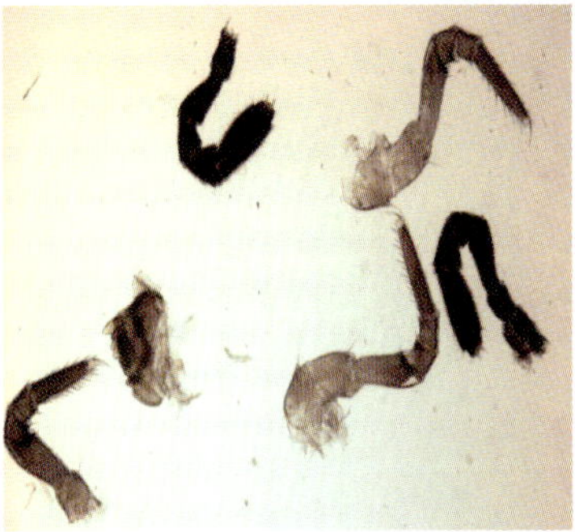

Millipede legs (x15 magnification)

Terrestrial woodlice *(Isopoda)*

- Eaten by all shrews.
- Easily confused with *Asellus.*
- Legs have short spines along the edge. Final segment before the claw is shorter than in *Asellus*. Claw is pigmented (unlike Asellus).
- Antennae are thick with few segments.
- Carapace is thick, highly chitinised and relatively inflexible.

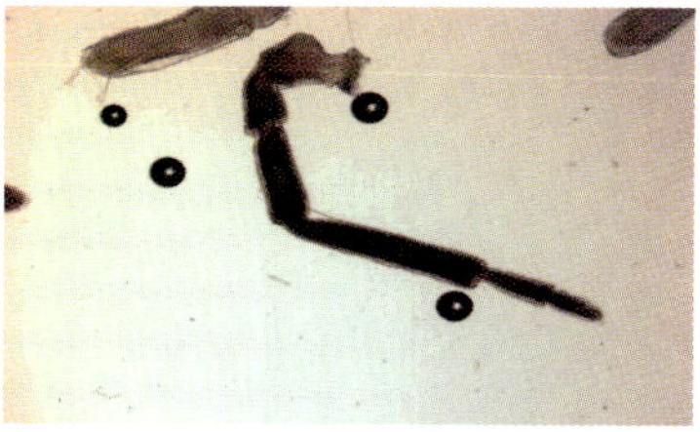

Woodlouse antennae (x12 magnification)

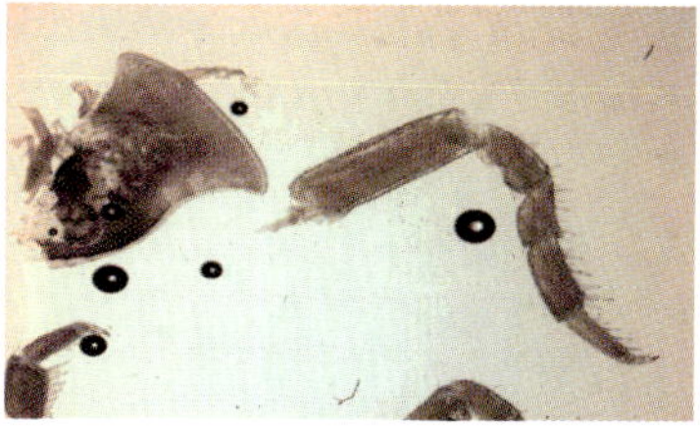

Woodlouse leg (right) and carapace (left) (x12 magnification)

Earthworms

- Eaten by water shrews and common shrews; rarely eaten by pygmy shrews.
- Chaetae are the only earthworm remains that will be found. Chaetae resemble small, yellowish shards of glass with a slight obtuse S-shape. They will be found lying on the bottom of the Petri dish.

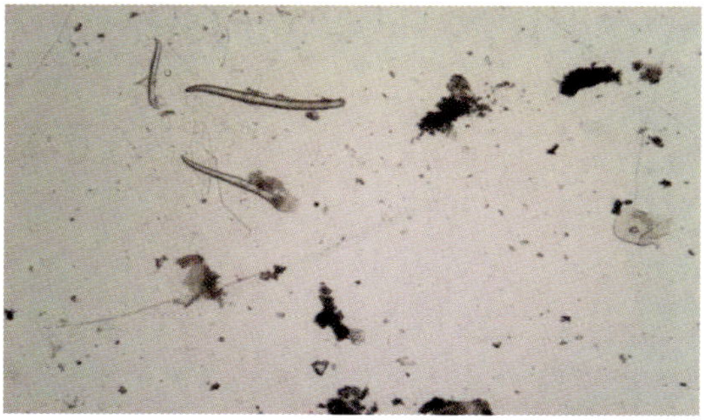

Earthworm chaetae (x80 magnification)

Molluscs

Both terrestrial and aquatic gastropod molluscs are eaten by water shrews but it is not possible to distinguish aquatic from terrestrial snails from the shell fragments found in scats.

This is not an exhaustive catalogue of all the invertebrate species eaten by water shrews. Finding evidence of aquatic prey in a scat generally indicates the presence of a water shrew at a site. On very rare occasions, common and pygmy shrews may consume aquatic prey if it is accessible from the bank. If the identification is unclear, use the colour and size of the scat as additional indicators of its origin, examine other scats or re-survey the area.

Appendix 3 – Health and safety information

Handling water shrews – Water shrews have a painful bite. Although their bites rarely break the skin and draw blood, they can induce slight reddening and inflammation at the site of the bite so it is advisable to wear gloves when handling water shrews. Any bites that do break the skin should be cleaned and covered with a plaster straight away.

Tetanus – Before commencing any fieldwork it is advisable to check that you are currently immunised against tetanus.

Leptospirosis (Weil's disease) – This is caused by a spirochaete which is excreted in the urine of wild rats and may occur in other rodents. Several species of *Leptospira* have been isolated from shrews. Avoid animals urinating on you, especially on to open wounds and cuts. Wash your hands thoroughly after handling any animals, live traps, bait tubes and hair tubes and after working in areas where rat urine may be present (*e.g.* canals, rivers, lakes and ponds). Make sure hands are clean before eating, drinking or smoking. Symptoms of the disease are much like influenza with headaches, fevers and chills developing within three to 19 days of exposure to the bacteria. Kidney problems and jaundice may occur in the later stages of the disease. If you have any symptoms after handling animals inform your doctor immediately. The earlier it is diagnosed and treated the better.

Lyme disease – This disease is also caused by a spirochaete which is transmitted in the bite of the tick, *Ixodes ricinus*. These blood-feeding ticks wait amongst vegetation in woodland, marshland and grassland for passing mammals and are particularly prevalent in areas where there are high numbers of deer. If you are working in these areas ensure that you have your trousers tucked into your socks and do not to wear shorts. Symptoms include the development of a rash or red patch which may be ring-shaped. Later symptoms are again flu-like and include stiff neck, fatigue and difficulty in concentrating. Consult your doctor immediately if you develop any of these symptoms after carrying out fieldwork. As with Weil's disease, the earlier the disease is treated the better.

Further health and safety information can be obtained from the Health and Safety Executive website www.hse.gov.uk and from The Mammal Society's website (www.mammal.org.uk).